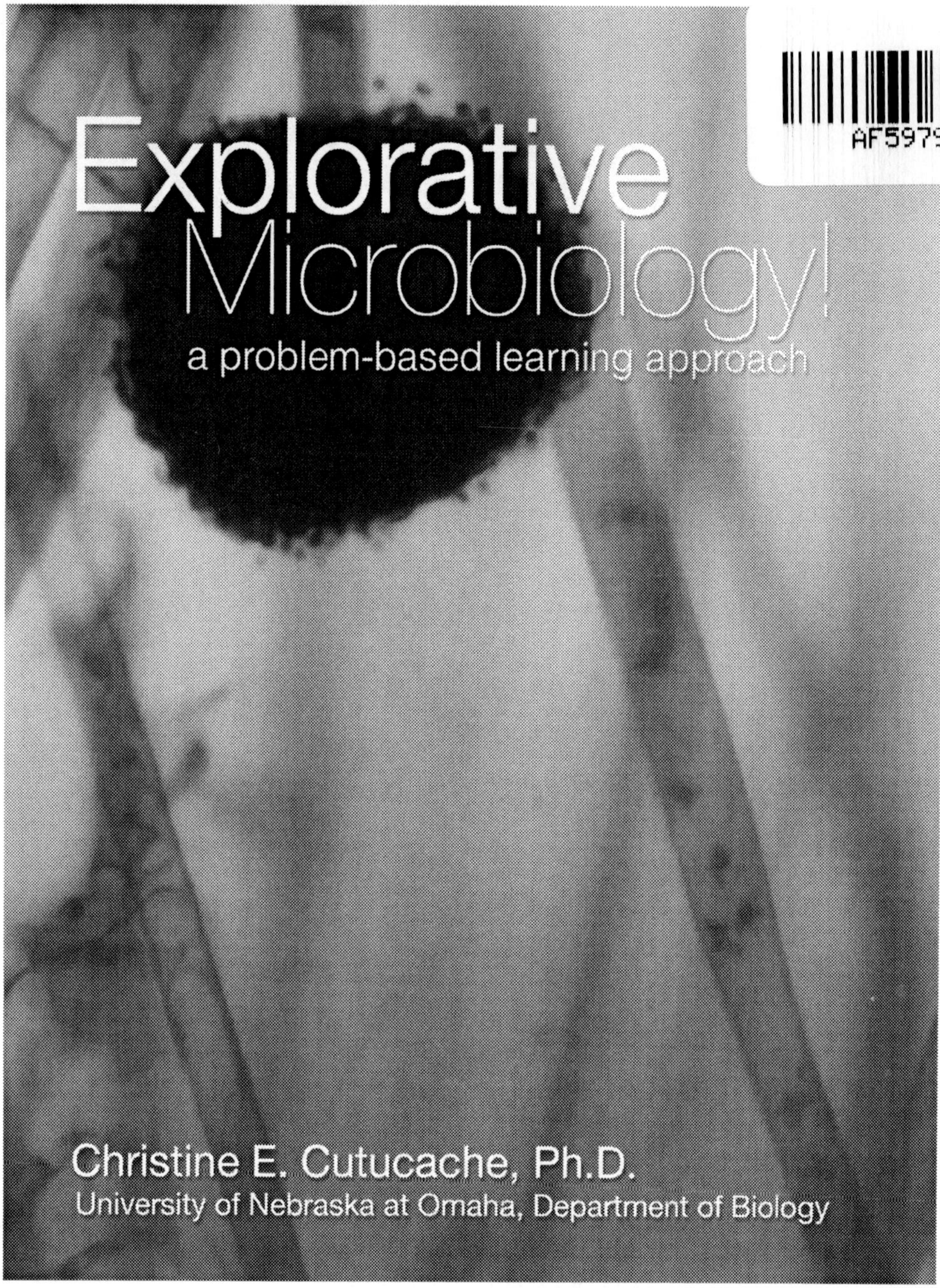

Explorative Microbiology!

a problem-based learning approach

Christine E. Cutucache, Ph.D.
University of Nebraska at Omaha, Department of Biology

Kendall Hunt
publishing company

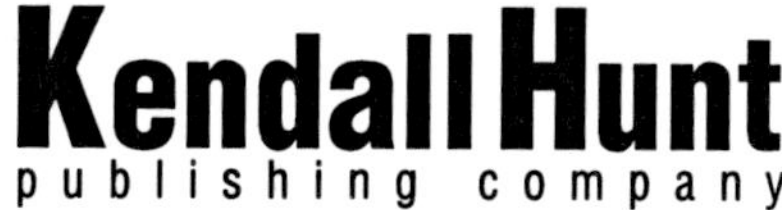

Cover design by LeDawna and James Strathman

Kendall Hunt
publishing company

www.kendallhunt.com
Send all inquiries to:
4050 Westmark Drive
Dubuque, IA 52004-1840

ISBN 978-1-4652-3991-4

Printed in the United States of America
10 9 8 7 6 5 4 3 2 1

Contents

Introduction

STUDENTS, WELCOME TO MICROBIOLOGY!

This booklet is intended to serve as a supplement to the material discussed in Microbiology and to open your eyes to the real-world, everyday applications of Microbiology. This book was written specifically for ***you***.

We will discuss the important molecular biological tools that scientists have harvested from microbes to solve crimes, cure diseases, and do basic research. We will also discuss how microorganisms influence the food you eat, the water you drink, and whether you can digest certain foods. Lastly, we will discuss emerging pathogenic microorganisms and cutting-edge technology that can be used for detection of these pathogens.

If you are a hands-on learner, this book will be your survival guide to Microbiology.

Have a great semester,

Dr. C

INSTRUCTIONS

Work in groups of 4 students. Within your group, assign the following roles:

- *Facilitator* (JOB: lead the group, ask discussion questions)
- *Time keeper* (JOB: make sure that your group stays within the allotted time period)
- *Recorder* (JOB: write down student input, including which idea came from whom to ask more later)
- *Remediator* (JOB: settle all of the disagreements, quarrels, & brawls)

If you have more than 4 students in one group, then the other students will be assigned the roles of "super awesome helpers."

EXPERIMENTATION USING THE SCIENTIFIC METHOD: A CASE OF THE MYSTERIOUS ILLNESS

This problem-based learning exercise corresponds with Chapters 1 & 2.

Instructions: With a group of 5 students, read through the script below. Think about the case at hand and the application of the scientific method. One group will be called at random to act out the situation with the rest of the class in order to solve the mystery illness. Be sure to entertain us.

Assign roles as to who will play the following: 1) Justin Bieber, 2) Justin's Grandmother, 3) the Physician (Doctor), 4) the Nurse, and 5) the Scientist, Dr. Schnazzy (Epidemiologist)

DIALOG:

(in the doctor's office)

Doctor: Good morning, I'm Doctor Fixalot. What is your name?

Beiber Fever: Hey Doc, I'm Justin Bieber.

Doctor: What seems to be the problem, Mr. Bieber?

Bieber Fever: Well, I've got this stomachache and I've had it a few days now. I took some Tums and Pepto and they haven't helped.

(enter nurse)

Nurse: Oh my gosh, oh my gosh, it's Justin Bieber!!!!! I am your biggest fan!

Beiber Fever: Well, thank you. I'll sign an autograph for you if you'd like?

Nurse: Yes, please! So, Doctor, what seems to be the problem?

Doctor: Well, Mr. Bieber seems to have a persistent stomachache that began a couple of days ago. Traditional over-the-counter medications aren't providing him any relief. Why don't you call in our epidemiologist, Dr. Schnazzy. You and Dr. Schnazzy can take a brief history while I go check on Mrs. Hickory in the next room.

(enter Dr. Schnazzy)

Dr. Schnazzy: Hello there, what seems to be the problem?

Nurse: Well, Justin has a stomachache and based on that, the Doctor suggested we take a brief history to see what might have caused it.

Dr. Schnazzy: Okay, that's a great idea. Mr. Bieber, could you retrace your steps the past couple of days? What have you had to eat?

Bieber Fever:

Well, 2 days ago I had:
Breakfast: cereal and milk
Lunch: pizza on the tour bus
Dinner: dinner at Grandmother's house (burger, fries, and ice cream)

And yesterday I had:
Breakfast: cereal and milk
Lunch: more pizza on the tour bus
Dinner: dinner with my fan club at Hell's Kitchen (porkchops, corn, and rice)

(simultaneously) *Dr. Schnazzy*: Pork! *Nurse*: Beef!

Bieber Fever: Uhhhh?

Dr. Schnazzy: Well you see, if meat is undercooked, it might contain bacteria and parasites that when ingested, set up a nice little oasis in your body and lead to illness. Do you remember if one of the two was a little pink in the middle when you bit into the meat?

Bieber Fever: (daydreaming…)

(Enter Grandma Bieber)

Grandma Bieber: Hi sweetie, what would you like for dinner?

Bieber Fever: I want one of your famous hamburgers! How do you make them taste so good?!

Grandma Bieber: Well, the trick is to cook them medium-rare to medium so they stay nice and juicy.

(awakens from daydream)

Bieber Fever: Okay, I remember. The pork at Hell's kitchen was cooked very well, I didn't see any pink in the center. When eating with Grandma, she said that she never cooks her burgers well done, but rather undercooks them to retain the good flavor!

Answer the following questions using the scientific method to track down the cause of Justin's illness.

1. List the steps in the scientific method.

2. Justin already *observed* his illness. Now he *questions* what the source was. Formulate a *hypothesis* statement for Justin.

3. The scientific method tells us that we must test our hypotheses using properly controlled experiments. Specifically, we need a group to which our data may be compared. Some experiments are easier to control than others. In the example above, what tests would you perform?

4. What controls would you use for the tests described in question 3?

5. As it turns out, the burger that Justin ate while at his Grandmother's house was not cooked thoroughly. Justin likely has food poisoning from *E. coli.* How could you prevent such an infection in the future?

6. Do you accept or reject your hypothesis statement?

The Causative Agent:

An Introduction to Laboratory Methods

(Corresponds to Chapter 3)

Often when someone is ill, we are unaware of the precise causative agent. Similarly, epidemiologists and pathologists (those that study disease incidence, prevalence, and infection) spend countless hours trying to determine the cause of a given infection or illness affecting a population. To determine the causative agent of any given illness, we can perform a variety of tests to identify it. This PBL exercise is intended to help you familiarize yourself with some preliminary laboratory techniques to study microorganisms. You will have an opportunity to explore more detailed techniques later in this exercise book.

SOLVE:

1. Describe the following steps to characterize a microorganism:
    1) Specimen collection
    2) Inoculation
    3) Incubation
    4) Isolation
    5) Inspection
    6) Identification

2. After collecting a soil sample, you identify 3 colonies growing on your agar plate. These colonies are all different in appearance thereby suggesting that they are in fact 3 unique species. Describe how you would determine which species these are and their biology.

3. List 3 common microorganisms found in soil and their potential pathogenicities.

Biotechnological Assays

Part I (DNA-based)

There exist endless biotechnological assays that we can use in the laboratory to identify a microorganism. In Chapter 3 you studied generic laboratory techniques and today you will review DNA-based biotechnological assays for the identification of microorganisms. Southern blotting, polymerase chain reaction (PCR), quantitative real-time PCR, and gel electrophoresis are all technologies available for DNA detection of microorganisms.

You gather samples of the microbiota at a nearby hot spring and after culturing your sample in the laboratory you observe a type of bacteria that you have never seen before. You show your results to your research mentor and he says that it looks like a novel species. To be certain that this is not a previously identified bacterium, you decide to determine the genome of this species.

SOLVE:

1. Describe the types of microorganisms that you would expect to find in a hot spring environment.

2. Describe 3 ways in which you could determine the genome of your bacterium of interest. Be sure to describe each step for each procedure from DNA purification through interpretation of your results. Discuss as a group and make sure everyone agrees on the protocols before presenting.

3. What are "molecular scissors" that bacteria possess that are helpful for the construction of vectors, running of DNA through gels, and clipping DNA?

Biotechnological Assays

Part II (RNA- and protein-based)

There are many biotechnological assays that we can use in the laboratory to study RNA. Today you will review RNA-based biotechnological assays. Northern blotting, polymerase chain reaction, gel electrophoresis, SDS-PAGE gels, and western blotting are some of the technologies available for RNA and protein detection. Therefore, we can study the RNA and protein expression in any nucleated cell.

Review Chapter 3 in your textbook on the procedures described above. Next, take turns in your group describing these in great detail. Make sure that all of the group members understand the procedures before moving on with the exercise.

SOLVE:

Sally arrives at the clinic and does not know what type of illness from which she is suffering. Her symptoms include night sweats, cough, and lethargy. Upon obtaining a history from Sally, you learn that she has had the common cold 6 times already this year. Additionally, Sally is a recovering alcoholic and she informs you that she used drugs about 10 years prior.

1. What diseases could Sally have based on the symptoms described? Is this likely a bacterium, parasite, or virus causing the infection?

2. Describe the tests that you would run in order to effectively diagnose Sally. (List 3 tests including a detailed description of the purpose behind each and how you would perform each test.)

3,. Based on how you answered questions 1 and 2, postulate what single disease Sally has. Furthermore, describe which tests would be positive informing you of this diagnosis.

4. Compare and contrast SDS-PAGE gels with that of agarose gels. Which are each used for?

Differential Media:

Who goes where? In what? With who?

(Chapter 3)

We can use nucleic acid- and protein-based methods to identify microorganisms, but we can also identify them based on their metabolic traits. For example, *E. coli* produce a unique metallic green color when grown on an Eosin Methylene Blue (EMB) plate making identification of this bacterium fast and easy. Other organisms are more difficult to identify and we must perform a variety of metabolic tests using differential media to determine which species are present. Chapter 3 of your textbook reviews "Tools of the Laboratory" including differential media types and the potential results from each. Use your knowledge of differential media to answer the following problem.

Your instructor noticed some slimy microorganism growing on the sinks in the bathroom at your school. She decided to collect the specimen and culture it in an effort to identify what it was. On an agar plate, the colony appeared yellow in color. After running a citrate and indole test, the results were negative. There was no growth on an EMB plate nor on a MacConkey (MAC) agar plate. This bacterium tested positively for having the enzyme catalase and yielded acid over acid (A/A) results in a triple sugar iron (TSI) slant.

SOLVE:

1. What are differential media? When would you use differential media?

2. What does the enzyme catalase do? Specifically, what are the end products?

3. From the results described above, which common bacterium do you think was growing in the sink in the bathroom?

4. Are *Staphylococus* species pathogenic 100% of the time? When are they pathogenic versus non-pathogenic?

5. A TSI slant can tell you many things about your bacterium of interest. If you saw growth only near the bottom of the slant and the color was red over yellow—what does this tell you about your microorganism?

The Case of the Unknown Emerging Disease

(Coincides with Chapter 4)

Joe just returned from a service-learning trip to South America. Upon returning, Joe gathered all of his friends to his home to regale the group of tales from his ventures.

Whilst Joe is telling the group of a hike he went on, Mary notices that Joe is very lethargic (which she reasons is a consequence of his elaborate travels) and has little sores all over his feet and lower leg (which she attributes to mosquito bites). Nonetheless, with great concern for her friend, coupled with her knowledge of world-wide pathogens, Mary decides to perform a literature search upon returning home for emerging infectious diseases in the region that Joe visited. Based on her analyses, she decides that based on Joe's symptoms, he likely has one of the following diseases: chromoblastomycosis, legionellosis, or schistosomiasis. Based on the information presented, answer the following questions.

SOLVE:

1. Compare and contrast "signs" with "symptoms." Provide 3 examples of each.

2. Does this infection most sound like a bacterial, fungal, parasitic, or viral infection?

3. Based on the case study presented above, provide your diagnosis for Joe and provide justification for this.

4. What are other signs and symptoms of this disease to watch for in coming days and weeks?

5. What is the name of the microorganism that causes this infection?

Parasites!

Neglected tropical diseases are a subset of diseases affecting tropical areas and claim many lives each year. These diseases are especially prevalent in lower-income regions of the world. Diseases on this list include Dengue haemorrhagic fever, Fascioliasis, Elephantiasis, Leishmaniasis, Chagas disease, Leprosy, and African Trypanosomiasis (African sleeping sickness). Schistosomiasis (*Schistosoma mansoni, S. haematobium,* or *S. japonicum*) affects 200 million people worldwide (CDC.gov). The parasitic infection is caused by worms that live in snails, so this disease is appropriately nicknamed, "Snail Fever."

Based on the above information about Schistosomiasis and what you have learned in class regarding anti-microbial drug therapy, answer the following questions.

SOLVE:

1. Antibiotics are very inexpensive and are easily administered (i.e., oral route the majority of the time). Would you recommend treatment of Schistosomiasis with an antibiotic? Why or why not?

2. Similarly to other parasitic infections, Schistosomiasis has a complex life cycle. In fact, it alternates between humans and snails found in freshwater. The infection of Schisto is from contact with contaminated freshwater. Draw an illustration of the life cycle of this parasite.

3. Based on your drawing in question 2, what is the vector and what is the host for Schisto?

4. Based on your drawing in question 2, do the worms or eggs lead to disease?

5. How could you prevent infection of Schisto?

Cestodes: *Taenia solium* (Flatworms)

Helminths are parasitic worms that exist all over the world. *Taenia solium* is a pork tapeworm that infects both pigs and humans and produces only mild abdominal symptoms. Prior to the development of tapeworms, which can take up to 3 months, this disease is unidentifiable by stool sample. However, antibody-capture methods work well to identify this parasite earlier. The illness caused by *T. solium* is cysticercosis. This disease causes seizures as it can infect the brain, subcutaneous tissue, eye, or liver. Treatment with Praziquantel is effective (CDC.gov).

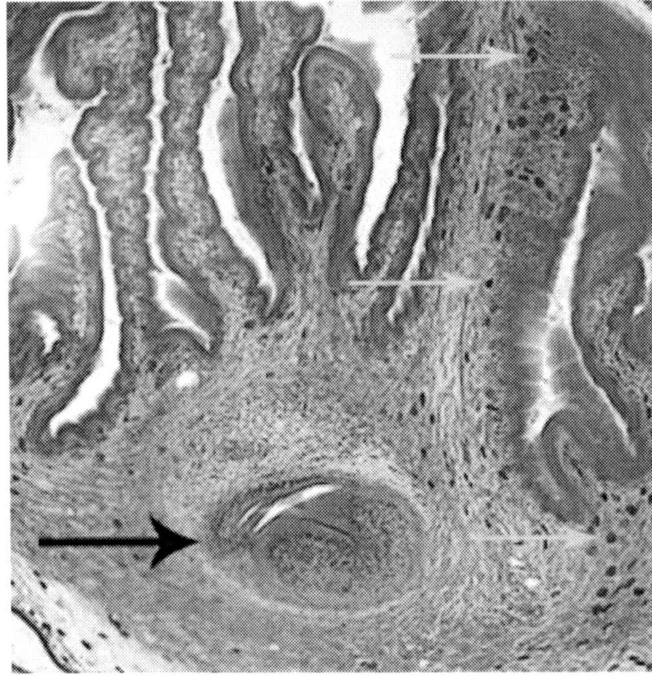

Photo courtesy of the CDC, 2013 (http://www.dpd.cdc.gov/DPDx/HTML/ImageLibrary/Cysticercosis_il.htm). Hematoxylin and eosin stain of a brain affected by Cysticercosis.

SOLVE:

1. Draw the life cycle of *Taenia solium.*

2. What other microorganisms have you learned about that are similar in life cycle and infectivity to *Taenia solium*?

3,. Describe 2 methods for antibody-capture. How could these be used to identify presence of *T. solium*?

4. What are 2 major, potential side effects with drug treatments to any infection?

Chagas Disease

(Chapters 5 & 13)

Trypanosoma cruzi is the parasite that causes Chagas disease (WHO). Approximately 8 million people worldwide are infected with this parasite with the highest incidence in Latin America. Individuals that are chronically infected can develop cardiac alterations. Triatomine bugs (shown below), aka "kissing bugs," serve as the vector for this disease. Triatomine bugs are active at night and use human blood as a food source. Therefore, when the kissing bugs bite a human for a blood meal, the exchange of feces from the bug transmits *T. cruzi* (WHO & CDC, 2013). Additionally, the parasite can be transmitted from mother to child or via organ donation. Therefore, vector control is the most effective method of disease prevention.

Although this disease is a tremendous financial burden on affected regions due to hospitalization and medical care, the cost for insecticides to prevent infection are more than double that cost (WHO, 2013). Chagas disease has been reported in the United States, Canada, Europe, and Western Pacific countries.

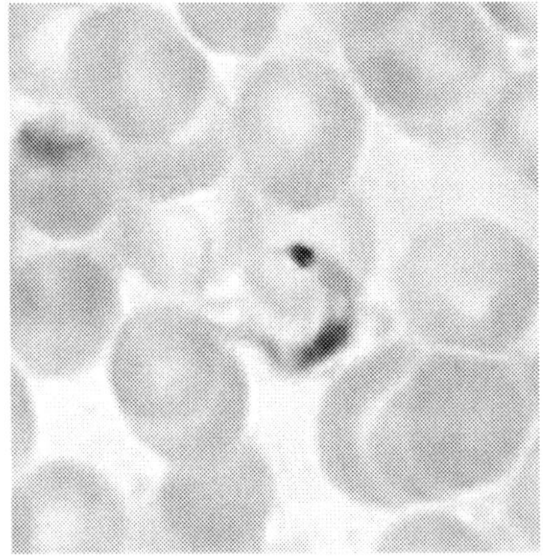

Photo courtesy of the CDC, 2013 (http://www.cdc.gov/parasites/chagas/diagnosis.html). Wright Giemsa stain of peripheral blood smear with *T. cruzi* at the center.

SOLVE:

1. Compare and contrast disease prevalence with disease incidence.

2. What are the pros and cons with treating disease symptoms rather than facilitating disease prevention?

3. Based on the epidemiology of Chagas disease, how could this disease be transmitted vertically (i.e., from mother to child)? Are any other diseases spread this way?

Emerging, Drug-resistant Infections:

Carbapenem-resistant Enterobacteriaceae (CRE)

On March 15, 2013, the Centers for Disease Control (CDC) published an announcement about an emerging hospital-associated infection. Carbapenem-resistant Enterobacteriaceae, or CRE, are being identified in clinical settings more commonly now. Current estimates of the mortality rate associated with a CRE infection lead to death in 50% of cases (cdc.gov).

Generally CRE infections occur as an opportunistic infection in the hospital while a patient is being treated for something different. Patients who use catheters, intravenous catheters, or those on ventilators are most susceptible (http://www.cdc.gov/hai/organisms/cre/).

Example of CRE organisms include *Klebsiella* species and *E. coli.* Although *E. coli* are normal flora, they can become resistant to multiple antibiotics.

SOLVE:

1. What is another name for a hospital-acquired infection?

2. What is an "opportunistic infection"? How is that different than a normal infection?

3. If *E. coli* are part of our normal, gut flora, how can they become resistant to multiple antibiotics? Provide 2 methods.

4. What are carbapenem antibiotics? (http://www.bmj.com/content/344/bmj.e3236 for additional information if needed)

5. Initially, cephalosporins were used to treat infections caused by Enterobacteriacea, why are carbapenems used now?

Epidemiology

(Corresponds to Chapter 13)

Illnesses come in all shapes and sizes. That is, they may affect a small cohort of individuals, city- or state-wide; they might affect an entire region, or even the world. Epidemiologists are those that study disease transmission, incidence, and prevalence. In essence, they "keep track" of how diseases spread and when and where they occur most frequently.

When a new disease emerges, epidemiologists and other scientists scramble to identify what it is, where it came from, its biology, and postulate how to eliminate it. For this exercise, you are an epidemiologist. There is a recent outbreak of Q Fever. The following information from the CDC describes Q Fever. Please discuss and answer the questions below based on your knowledge of epidemiology and microbiology.

The bacterium *Coxiella burnetii* causes a disease known as "Q Fever." In brief, the reservoirs for this disease are herding animals: cattle, sheep, and goats (http://www.cdc.gov/qfever/). *C. burnetii* are excreted in milk, urine, and feces thereby providing an easy method of transmission from organism to organism. *C. burnetii* is resistant to heat-drying and common disinfectants. Which other microorganisms have you learned about thus far with similar properties? *C. burnetii* infects humans by inhalation. Symptoms include high fever, severe headache, malaise, chills and/or nightsweats, vomiting, chest & abdominal pain (http://www.cdc.gov/qfever/) present approximately 2 weeks after exposure. For more information regarding chronic Q Fever, please visit the CDC's website.

SOLVE:

1. As an epidemiologist, describe how you would identify whether someone has Q Fever (include lab & non-lab techniques in your answer).

2. Recently, you notice that a city in rural Wisconsin has a 60% incidence of Q Fever. You decide to travel to the region to investigate the localized epidemic. Postulate the most likely cause(s) of such a localized infection and describe some of the tests that you would run upon arriving there.

3. Use the following equations to calculate the prevalence and incidence of Q Fever in Wisconsin.

   Prevalence = (Total number of cases in population) / (Total number of persons in population) x 100

   Incidence = (Number of new cases) / (Total number of susceptible persons) [report per 100,000 persons]

4. Compare and contrast disease incidence with disease prevalence. Provide a specific example of each regarding Q Fever.